美好的生活

失败与想象力不可或缺

〔英〕J.K.罗琳 著 〔美〕乔尔·霍兰 绘 任战 译

人民文学出版社
PEOPLE'S LITERATURE PUBLISHING HOUSE

著作权合同登记号：图字 01-2017-622

图书在版编目（CIP）数据

美好的生活：失败与想象力不可或缺 / (英) J.K.罗琳著；(美) 乔尔·霍兰绘；任战译. -- 北京：人民文学出版社, 2017
ISBN 978-7-02-013453-3

Ⅰ. ①美… Ⅱ. ①J… ②乔… ③任… Ⅲ. ①散文集－英国－现代 Ⅳ. ①I561.65

中国版本图书馆CIP数据核字(2017)第255201号

责任编辑 甘 慧 张玉贞
装帧设计 汪佳诗

出版发行 人民文学出版社
社　　址 北京市朝内大街166号
邮政编码 100705
网　　址 http://www.rw-cn.com

印　　制 上海盛通时代印刷有限公司
经　　销 全国新华书店等

字　　数 50千字
开　　本 890毫米×1240毫米 1/32
印　　张 4.625
版　　次 2019年1月北京第1版
印　　次 2019年1月第1次印刷

书　　号 978-7-02-013453-3
定　　价 39.90元

如有印装质量问题，请与本社图书销售中心调换。电话：010-65233595

美好的生活

福斯特校长、哈佛理事会和监事会的各位成员、各位教职员工、自豪的父母们，还有，最重要的，各位毕业生们：

Thank
You

我想说的第一句话是“谢谢你们”。不仅是因为哈佛大学给了我这份殊荣，还因为一想到要在这里做毕业致辞让我数周来一直担忧、紧张，以致体重都轻了一些。这真是一个双赢的局面！我现在要做的就是深呼吸，偷瞄几眼红色的旗帜，假装我正身处世界上最大规模的格兰芬多学院的聚会上。

毕业致辞是一份巨大的责任，或者说我是这么认为的，直到我的思绪回到了自己的毕业典礼。那天的致辞嘉宾是英国著名哲学家玛丽·沃诺克女男爵。回忆她的演讲对我写今天的发言稿帮助很大，因为我发现，自己一个字也想不起来她到底

讲了什么。这一发现令我如释重负，让我不再担心无意中会对你们施加的影响，使你们放弃在商业、法律和政治领域的大好前程，而去当一名快乐的巫师。

你们看，如果多年以后你们都还记得这个“快乐巫师”的笑话，那就说明我的演讲比玛丽·沃诺克女男爵成功。确定可以实现的目标，这是自我完善的第一步。

事实上，我耗费心力、绞尽脑汁地思考今天该对你们说些什么。我问自己，有哪些事情是我希望当年毕业时就该知道的，还有哪些重要的事情，是我从毕业那天到现在的二十一年间学到的。

The Importance of IMAGINATION

我找到了两个答案。在今天这样一个美好的日子里，当我们聚在一起，庆祝你们的学术成就时，我决定跟你们谈一谈失败的益处。另外，正因你们站在所谓的“真实人生”的门槛上，我想要赞颂想象力的至关重要性。

UNEASY

这两个话题似乎是异想天开或是自相矛盾的，但请耐心听我说。

对于四十二岁的我来说，回顾毕业典礼上二十一岁的自己并不是那么愉快的。

二十一年是我人生的一半，而半生前的我正艰难地在自己的理想和家人的期待之间谋求平衡。

BALANCE

PERSONAL QUIRK

当时的我深信，我唯一想做的就是写小说，永远都是。而我的父母，他们都出身贫寒，都没有上过大学，在他们看来，我过度活跃的想象力不过是种可笑的个人怪癖，靠它是永远没法支付按揭贷款或是老有所依的。好吧，我知道如今看来这件事有多讽刺。

所以，父母希望我能够去拿一个有职业前景的学位，而我想学英国文学，于是我们各让一步，达成了妥协——我去学当代语言。后来看来，妥协的结果让每个人都不满意。还没等父母的车拐过马路尽头的街角，我就扔掉德语，一头扎进了古典文学的长廊。

我记得自己并没有告诉父母我在学习古典文学，他们可能是在我毕业那天才知道的。我想，在这个星球上的所有学科之中，他们大概找不出比希腊神话对于获得豪华卫生间的钥匙来说更没用的科目了。

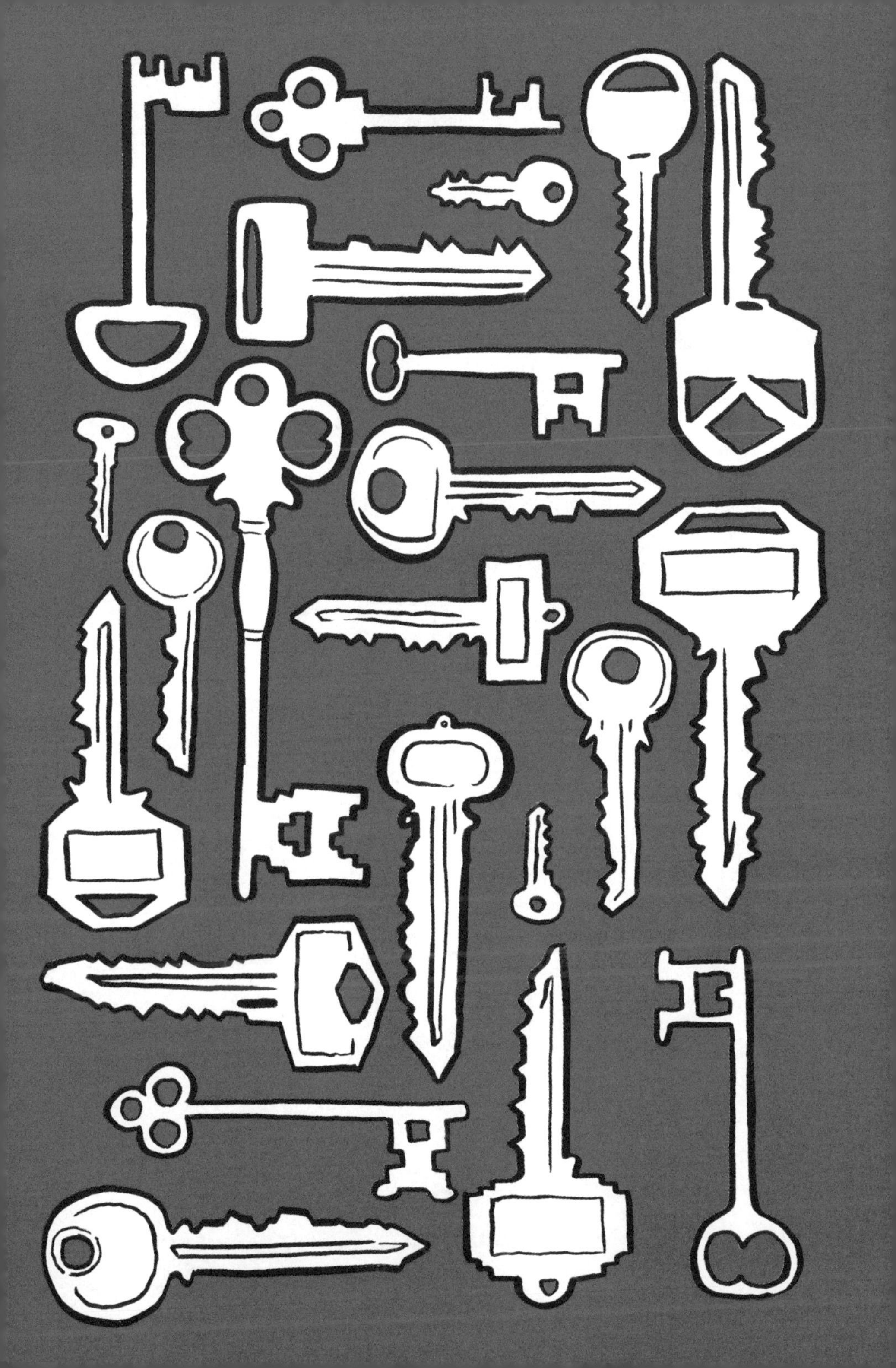

插一句，我想阐明一点，那就是我并不因为父母的观念而心怀怨气。终有一天，你会停止责怪父母当年错误的指引。当你长到能够取得驾照、自己掌握方向盘的年龄，就应该自己负起责任了。此外，我的父母希望我不再经历贫穷，我不能因为这样的心愿去怪他们。他们吃过穷日子的苦，我也穷过很长一段时间，我非常同意他们的看法，那就是，贫穷绝不是什么让人高贵的经历。贫穷带来忧虑、压力，有的时候还有抑郁；贫穷意味着无数次琐碎的窘迫和艰难。靠自身的奋斗摆脱贫穷是值得骄傲的，但只有愚蠢的人才会赋予贫穷浪漫的诗意。

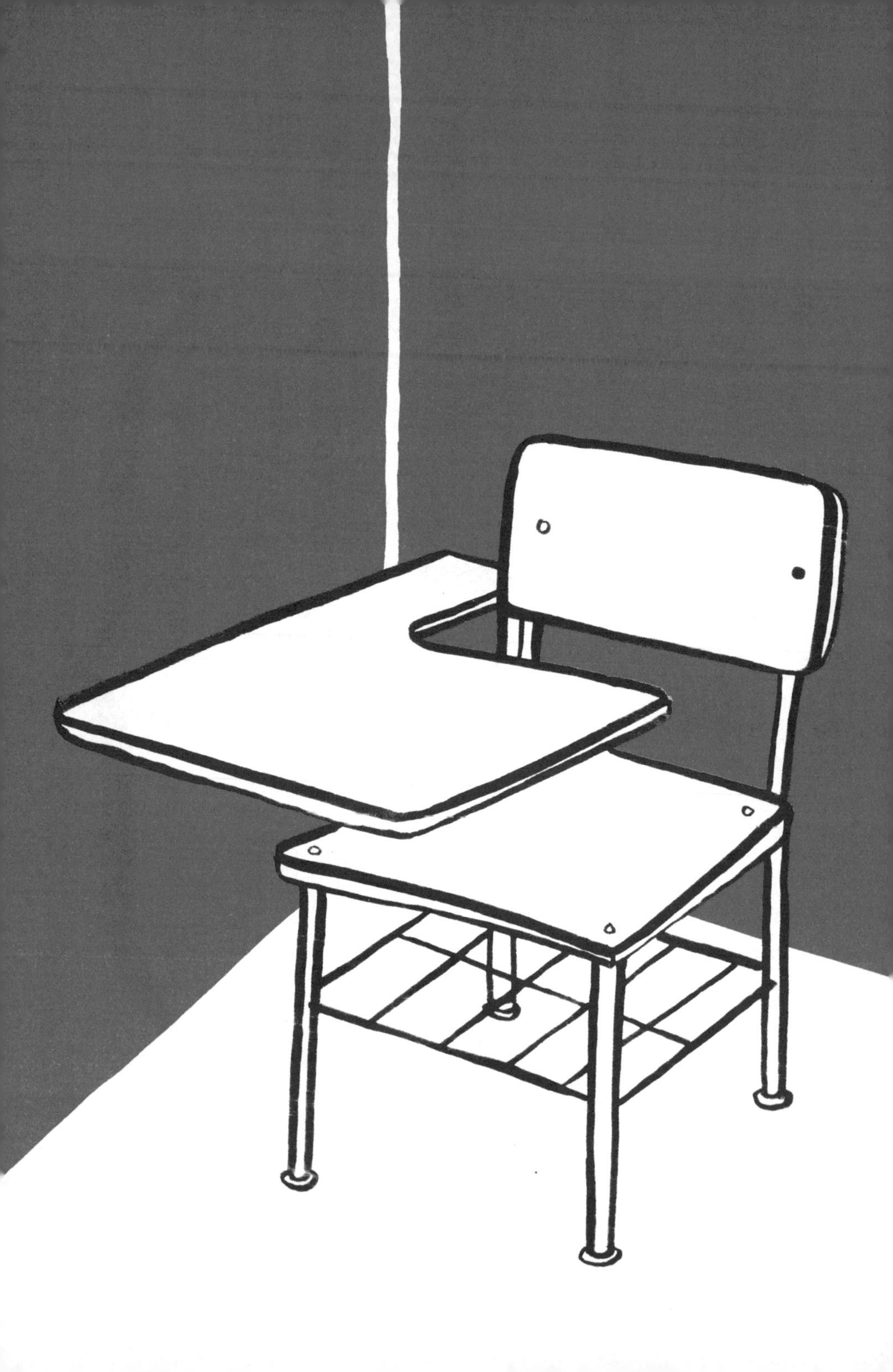

不过，在你们这个年龄的时候，我最怕的不是贫穷，而是失败。

在你们这个年龄的时候，虽然我在大学里缺乏明确的动力，花了太多的时间在咖啡厅里写故事、太少的时间去听课，但我对于怎么应付考试还是有窍门的，毕竟多年来，这是衡量我和我的同龄人成功与否的标准。

我不会笨到因为你们年轻、有才华、受过良好的教育，就会认为你们从来不知愁的滋味。天赋和智慧永远不能保护一

个人免于命运任性的摆布，而我也没有一秒钟认为在座各位的生活会是一帆风顺、无忧无虑的。

然而，毕业于哈佛大学这个事实就暗示了你们对于失败并不是很熟悉。这一路驱使你们的对失败的恐惧或许跟对成功的渴望一样多。甚至，你们眼中的失败可能跟普通人眼中的成功相差不远，毕竟你们已经飞了那么高。

SO HiGH
HAVE
YOU
ALREADY
FLOWN

"I was the biggest failure I knew"

最终，我们都会建立自己对于失败的定义，但如果你愿意，这个世界早已迫不及待地向你提供一套标准。所以我想，或许我可以说，按照任何传统标准，毕业仅仅七年的我，那是史诗级的失败了。我刚刚结束了一场异常短暂的婚姻，没有工作，独自抚养一个孩子，过着现代英国人所能想象的最贫穷的生活，只差流落街头了。父母对我的担忧，我对自己的担忧，都成为了现实。以每一条常规标准来判断，我都是我知道的最失败的人。

此刻，我站在这里，并不打算对你们说失败是有趣的。那是我人生中的黯淡时光，绝想不到它日后会变成媒体所描述的那种童话般奇妙的绝地反击。它就像黑暗的隧道，我不知道它会延伸到多远的地方，在很长一段日子里，隧道尽头的任何一点光对我来说都只是希望，而不是现实。

那么，我为何要谈失败的益处呢？仅仅是因为，失败意味着剥离任何无关紧要的东西。我不再自欺欺人，假装成别人的样子，而是开始将所有的精力都投入到对我来说唯一重要的工作上。若是我之前做成了别的事情，我或许永远都不会下决心

在我真正有归属感的领域努力取得成功。

我感觉自己获得了自由，因为最害怕的事情都已经成真，而我还活着。我不仅活着，还拥有一个可爱的女儿、一台老打字机和一个大构思。就这样，人生的谷底变成了坚实的根基，我在上面重建生活。

你们也许永远不会像我一样输得那么惨，但生活中，失败总是难免。没有人可以在任何事情上都不失败，除非他活得战战兢兢、谨小慎微，但那样并不叫真正的活着。换个说法就是，人无论如何都将面对失败。

失败给了我内在的安全感，这种安全感是任何考试合格都带不来的。失败教我认识自己，这种认识是通过其他方式无法获得的。我发现我比自己想象中更意志坚强、克己自律，还发现我有那么好的朋友，他们的价值完全胜过宝石。

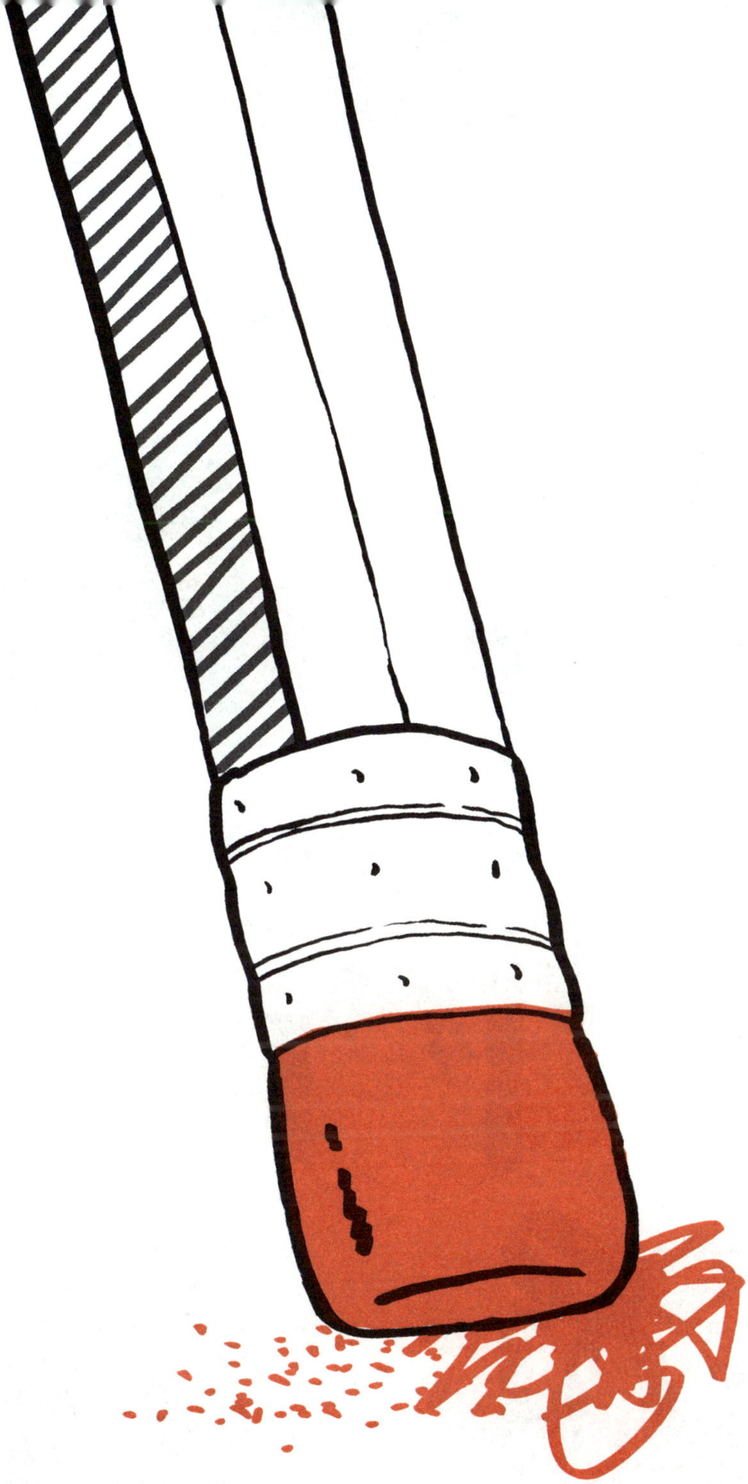

意识到自己经历挫折之后更明智、更强大，就意味着你从此以后拥有了克服困难的能力。若非经过逆境的考验，你不会真正认识到自我，或者亲情、友情、爱情的强韧。这样的认识是真正的财富，因为它们得来不易，也比我获得的任何学历证书都更有价值。

humility

所以，如果有时间转换器，我会告诉二十一岁的自己，只有明白人生不是一张知识清单或事业成就清单，人才能够幸福。你们的学位、履历，并不是你们的人生，尽管你们以后会碰到很多将二者混淆的人，他们和我同龄甚至比我年长。生活是艰难的、复杂的，而且不受任何人完全掌控。谦卑地承认这一点，我们才能笑对世事无常。

你们或许会认为，我选择今天演讲的第二个主题——想象力的重要性，是因为它在我重建生活的过程中扮演了重要的角色，但事实并非完全如此。尽管我个人会不遗余力地捍卫睡前故事的价值，但我已经学会珍视更广义层面上的想象力。想象力不仅是人类特有的能够预见本不存在之物的能力，因此是一切发明和创新的源头；想象力也有可能是最具转换力和启发性的能力，可以让我们对并未亲身体验过的他人的经历感同身受。

对我的人生影响最大的经历之一发生在《哈利·波特》之前，它教会我的很多东西，我后来写进了《哈利·波特》系列。那是我最早的全职工作之一的意外收获。那段时间，虽然我会在午休时候开小差写故事，但我二十出头时的房租是靠我在大赦国际伦敦总部的非洲研究部工作的工资来支付的。

我在我的小办公室里匆匆读着那些字迹潦草的书信，写信人冒着坐牢的风险，将这些信偷运出国，只为揭露极权统治的真相。我看见了由绝望的亲友寄来的那些失踪者的照片。我阅读着被迫害者的证词，看着他们伤口的照片。我打开一份份目击证人手写的记录，关于即决审理和行刑，关于绑架和强奸。

我的许多同事以前是“政治犯”，他们被人从家中带走，或是被迫流亡，只因为他们胆敢发表了针对政府的反对言论。造访我们办公室的人们，有的来提供信息，有的来打听留在国内的亲友的消息。

我永远也不会忘记一位来自非洲的曾被严刑拷打的受害者，他跟我年龄相仿。有过那些不幸的经历之后，他的精神状况不太好。对着摄像机讲述发生在自己身上的暴行时，他的身体会不受控制地发

抖。他比我还高三十厘米，但看上去像孩子般脆弱。谈话结束后，我被安排送他去地铁站。这个被暴行摧毁了人生的男人极有礼貌地握住我的手，祝我幸福。

A
SCREAM
OF
PAIN
AND
HORROR

有生之年，我会永远记得我走在空荡荡的走廊上，突然听到从一扇关着的门后传来尖叫，那叫声充满痛苦、恐惧，我此后再未听过。那扇门开了，研究员探出头来，让我赶快去给坐在她身边的那个年轻人拿一杯热饮来。她刚刚告诉年轻人，为了报复他发布的反对言论，极权政府逮捕并处决了他的母亲。

那时我二十出头，工作的每一天都在提醒我自己有多么幸运，生在一个民主选举的国家，获律师帮助和公开审判是每一个人的权利。

每一天，我都看到更多证据，看到人类为了获得或维持权力而对自己的同胞犯下的罪行。我开始做噩梦，真的是噩梦，梦到我看到、听到和读到的东西。

然而，也是在大赦国际，我比以往更多地看到了人性的善良。

大赦国际动员了成千上万没有因自己的信仰而被迫害和囚禁的人们去帮助那些有过此类经历的人。人类同情心的力量促进了挽救生命、释放囚徒的集体行动。有许许多多个人福祉和安全都有保障的普通人团结协作，去拯救那些素不相识也永远不会遇见的人。我微不足道的参与是我一生中最光荣，也最有感悟性的经历之一。

与地球上的其他生物不同，人类不用亲身经历，就可以了解和理解事物。通过想象，他们就能够跟别人易地而处。

当然，想象力是一种能力，就像我书中写到的魔法那样，在道德上是中立的。人可以用这种能力去操纵或控制，也可以去理解或同情。

"They can think themselves into other people's places"

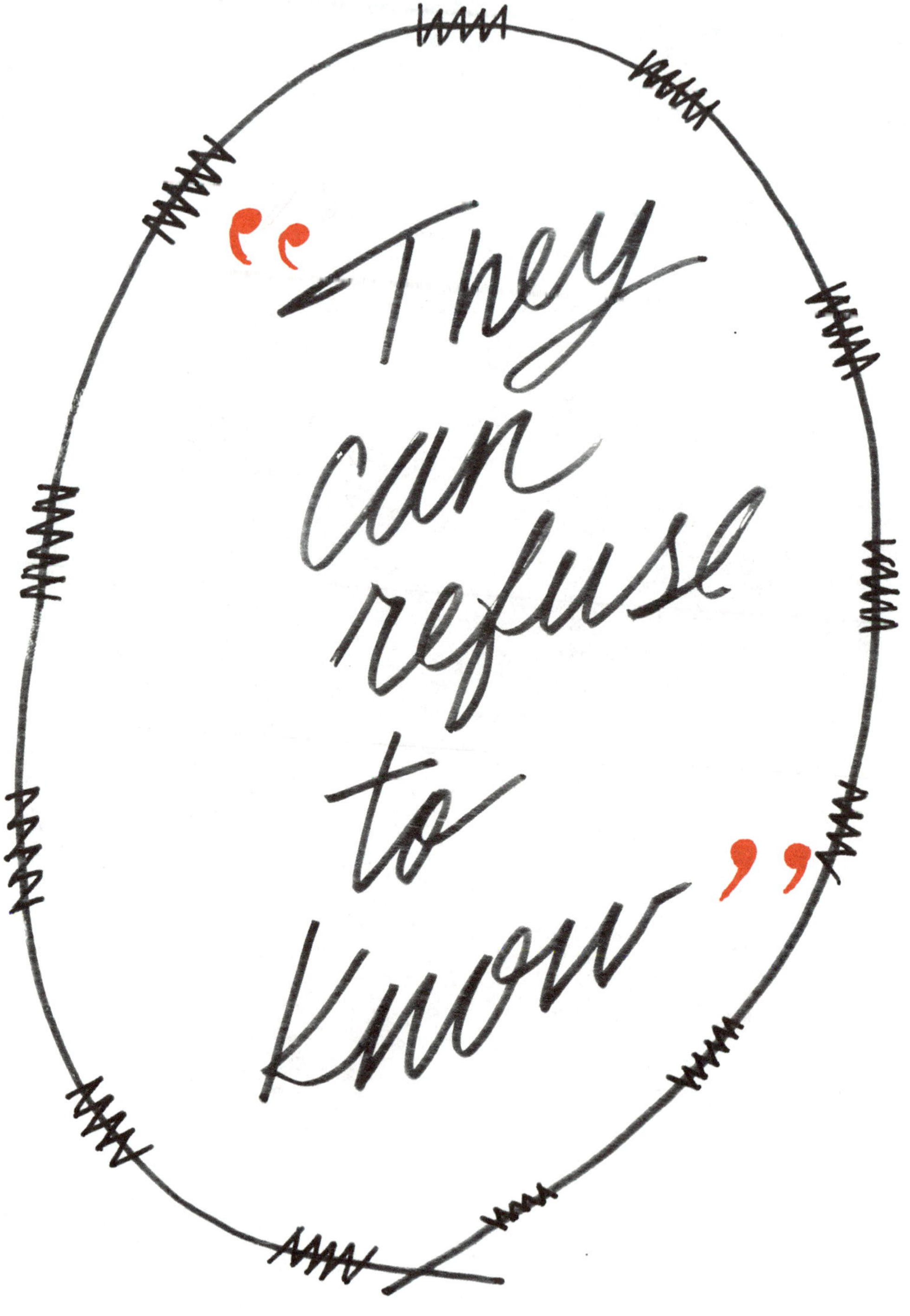
"They can refuse to know"

还有很多人根本不去运用他们的想象力。他们选择待在自身生活的舒适区，从不费神去想，若是生在不同的环境下会是什么感觉。他们可以拒绝听见哀号，他们可以拒绝看见牢笼；只要痛苦不触及自身，他们就关闭自己的头脑和心灵；他们可以拒绝知情。

或许那样的生活方式也有其吸引力，但我并不觉得他们做的噩梦会比我少。选择住在狭窄之地会导致一种精神性的广场恐惧症，它自有其恐怖之处。我想，刻意关闭想象力的人会看见更多恶魔，他们

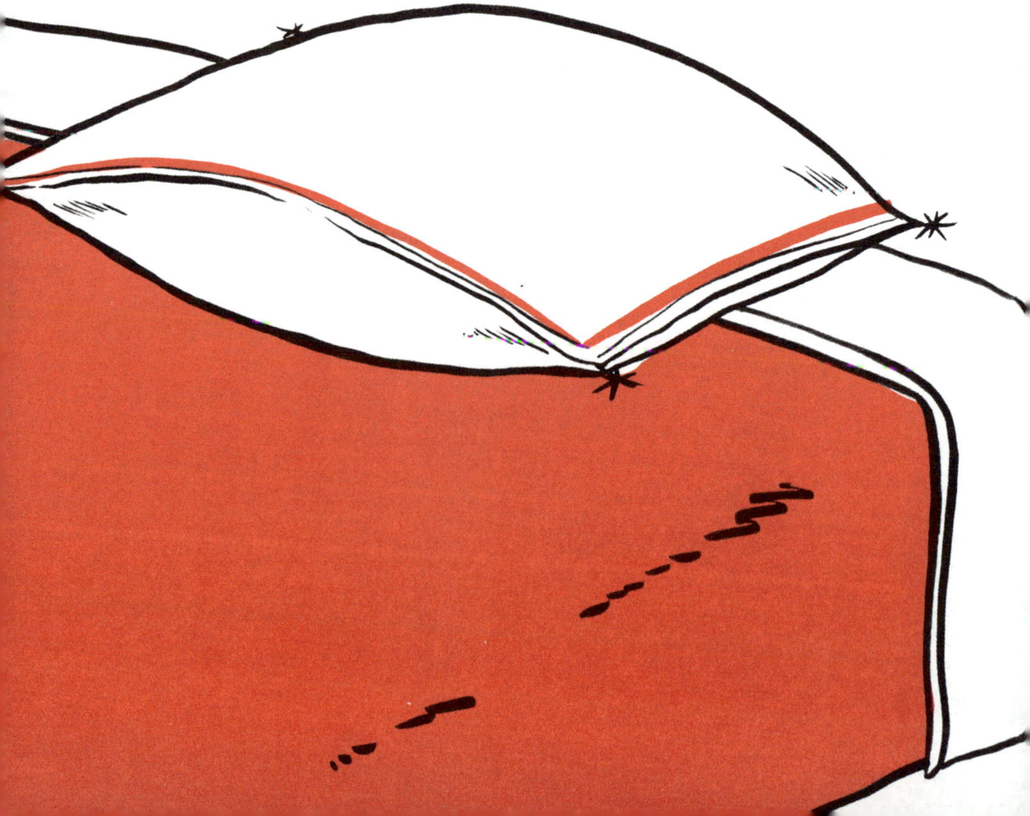

通常更胆怯。

更有甚者，选择漠视的人会成就真正的恶魔。因为，尽管没有亲手作恶，我们却通过自己的冷漠成了“恶”的帮凶。

十八岁的我在古典文学的长廊中摸索，寻找我当时并不能清晰定义的东西。收获之一是我看到了希腊哲学家普鲁塔克的一句话：“内在的实现会改变外在的现实。”

这句话振聋发聩，在我们的生活中上千次被证明是正确的。它部分地表达了我们与外部世界不可分割的联系，我们的存在本身，就可以影响他人的生活。

N
W
E
S

但是你们，哈佛大学2008年的毕业生们，对他人生活的影响又会有多大呢？你们的才华，你们的勤奋，你们争取并得到的教育机会，给了你们独特的身份和责任。即使是你们的国籍，也让你们与众不同。你们中的大多数，属于这个世界上仅存的超级大国。你们选举的方式，你们生活的方式，你们抗议的方式，你们对政府施加的压力，都会对远在国境之外的地方造成影响。这是你们的特权，也是你们的重担。

如果你们选择运用自己的身份和影响，去为沉默之人发声；如果你们选择不仅与权势之人、也与弱势群体站在一起；如果你们保有想象力，去理解那些生存条件不如你们优越的人们，那么庆幸与你们在一起的，将不只是你们自豪的家人，还有无数因为你们而改善了现状的人们。我们不需要魔法来改变世界，因为我们需要的所有力量本来就存在于自身：我们有能力去想象一个更好的世界。

我的演讲即将结束。我对你们还有一个最后的祝福，而这是我在二十一岁时已经拥有的。毕业日那天坐在我身边的那些朋友，是我一生的朋友。他们是我孩子的教父教母，他们是我在遇到困难时可以求助的人，就算我把他们的名字用在食死徒①的身上，他们也没有跟我翻脸。毕业之后，把我们连在一起的，有深厚的情谊，有共同拥有的、再也回不去的青春年华，当然，我们还握有某些照片铁证，如果我们中的某些人要竞选首相，那些照片将会价值连城。

---

①译注：食死徒（Death Eaters），是小说《哈利·波特》中的组织，伏地魔党羽的称号。这个组织的成员都是黑魔王的支持者和信徒。他们左臂上都被烙印上黑魔标记，而且精通黑魔法。

所以，今天，我祝福你们拥有同样的友情。明天，哪怕你们一个字也想不起来我说了些什么，我也希望你们能记住塞涅卡。他是我逃离职业阶梯躲进古典文学长廊、寻找古人智慧时遇到的一位古罗马哲人。他说了这样一句话：

“生活就像故事一样，不在其多长，而在其多好。”

我祝你们每个人都拥有美好的生活。谢谢你们。

I wish you all very good lives

Very Good Lives

President Faust, members of the Harvard Corporation and the Board of Overseers, members of the faculty, proud parents, and, above all, graduates.

Thank You

The first thing I would like to say is 'thank you'. Not only has Harvard given me an extraordinary honor, but the weeks of fear and nausea I have endured at the thought of giving this commencement address have made me lose weight. A win-win situation! Now all I have to do is take deep breaths, squint at the red banners, and convince myself that I am at the world's largest Gryffindor reunion.

Delivering a commencement address is a great responsibility, or so I thought until I cast my mind back to my own graduation. The commencement speaker that day was the distinguished British philosopher Baroness Mary Warnock. Reflecting on her speech has helped me enormously in writing this one,

because it turns out that I can't remember a single word she said. This liberating discovery enables me to proceed without any fear that I might inadvertently influence you to abandon promising careers in business, the law, or politics for the giddy delights of becoming a gay wizard.

You see? If all you remember in years to come is the gay wizard joke, I've come out ahead of Baroness Mary Warnock. Achievable goals: the first step to self-improvement.

Actually, I have racked my mind and heart for what I ought to say to you today. I have asked myself what I wish I had known at my own graduation, and what important lessons I have learned in the twenty-one years that have expired between that day and this.

The Importance of Imagination

I have come up with two answers. On this wonderful day when we are gathered together to celebrate your academic success, I have decided to talk to you about the benefits of failure. And as you stand on the threshold of what is sometimes called 'real life', I want to extol the crucial importance of imagination.

UNEASY

These may seem quixotic or paradoxical choices, but please bear with me.

Looking back at the twenty-one-year-old that I was at graduation is a slightly uncomfortable experience for the forty-two-year-old that she has become. Half my lifetime ago, I was striking an uneasy balance between the ambition I had for myself and what those closest to me expected of me.

PeRSoNAl
QuiRK

I was convinced that the only thing I wanted to do, ever, was write novels. However, my parents, both of whom came from impoverished backgrounds and neither of whom had been to college, took the view that my overactive imagination was an amusing personal quirk that would never pay a mortgage or secure a pension. I know that the irony strikes with the force of a cartoon anvil now.

So they hoped that I would take a vocational degree; I wanted to study English Literature. A compromise was reached that in retrospect satisfied nobody, and I went up to study Modern Languages. Hardly had my parents' car rounded the corner at the end of the road than I ditched German and scuttled off down the Classics corridor.

I cannot remember telling my parents that I was studying Classics; they might well have found out for the first time on graduation day. Of all the subjects on this planet, I think they would have been hard put to name one less useful than Greek mythology when it came to securing the keys to an executive bathroom.

I would like to make it clear, in parenthesis, that I do not blame my parents for their point of view. There is an expiry date on blaming your parents for steering you in the wrong direction; the moment you are old enough to take the wheel, responsibility lies with you. What is more, I cannot criticize my parents for hoping that I would never experience poverty. They had been poor themselves, and I have since been poor, and I quite agree with them that it is not an ennobling experience. Poverty entails fear, and stress, and sometimes depression; it means a thousand petty humiliations and hardships. Climbing out of poverty by your own efforts—that is something on which to pride yourself, but poverty itself is romanticized only by fools.

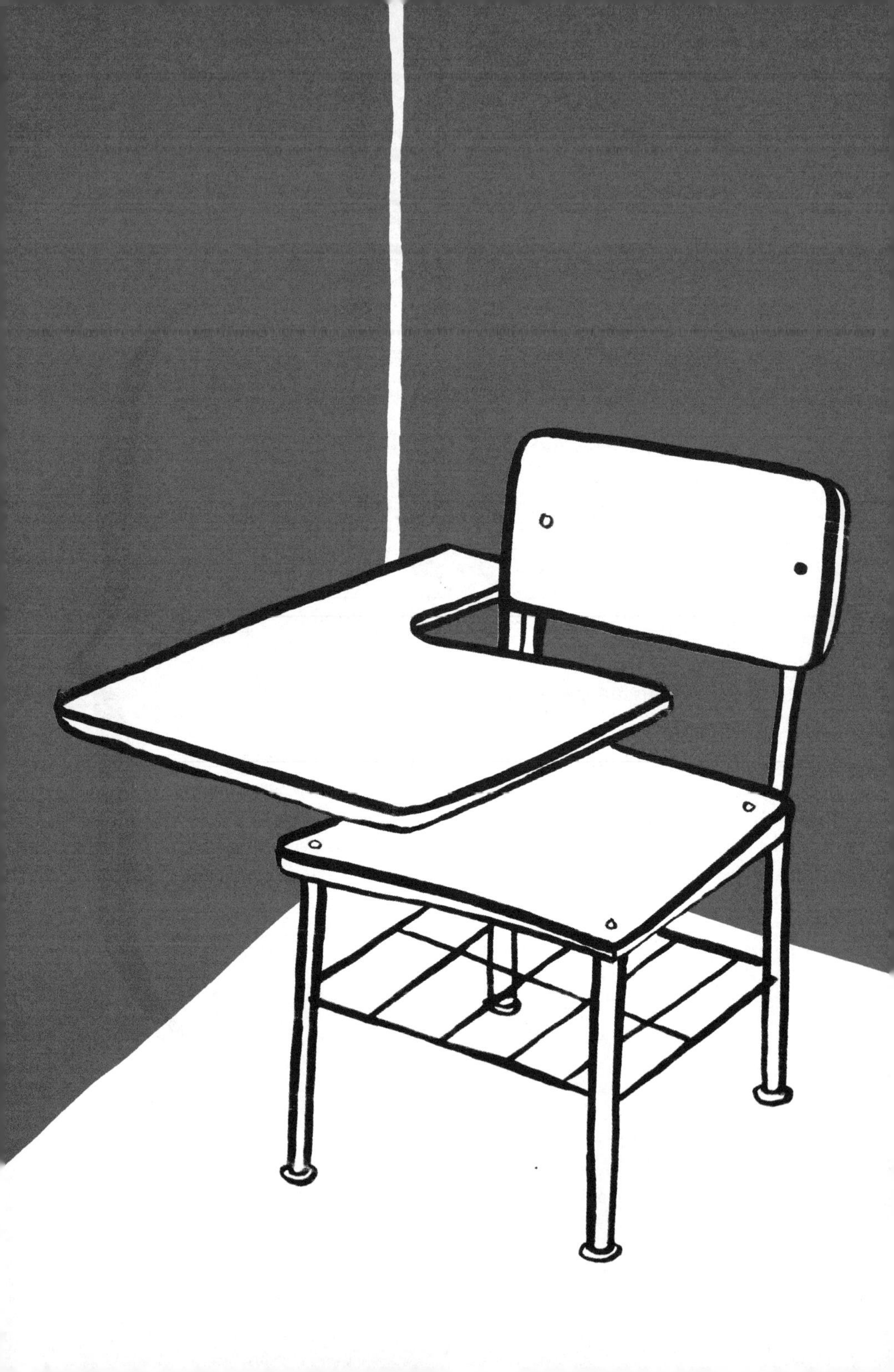

What I feared most for myself at your age was not poverty but failure.

At your age, in spite of a distinct lack of motivation at university, where I had spent far too long in the coffee bar writing stories and far too little time at lectures, I had a knack for passing examinations, and that, for years, had been the measure of success in my life and that of my peers.

I am not dull enough to suppose that because you are young, gifted, and well-educated, you have never known hardship or heartache. Talent and intelligence never yet inoculated anyone against the caprice of the Fates, and I do not for a moment suppose that everyone here has enjoyed an existence of unruffled privilege and contentment.

However, the fact that you are graduating from Harvard suggests that you are not very well acquainted with failure. You might be driven by a fear of failure quite as much as a desire for success. Indeed, your conception of failure might not be too far removed from the average person's idea of success, so high have you already flown.

SO HiGH
HAVE YOU
ALREADY
FLOWN

"I was the biggest failure I knew"

Ultimately we all have to decide for ourselves what constitutes failure, but the world is quite eager to give you a set of criteria, if you let it. So I think it fair to say that by any conventional measure, a mere seven years after my graduation day, I had failed on an epic scale. An exceptionally short-lived marriage had imploded, and I was jobless, a lone parent, and as poor as it is possible to be in modern Britain without being homeless. The fears that my parents had had for me, and that I had had for myself, had both come to pass, and by every usual standard I was the biggest failure I knew.

Now, I am not going to stand here and tell you that failure is fun. That period of my life was a dark one, and I had no idea that there was going to be what the press has since represented as a kind of fairy-tale resolution. I had no idea then how far the tunnel extended, and for a long time any light at the end of it was a hope rather than a reality.

So why do I talk about the benefits of failure? Simply because failure meant a stripping away of the inessential. I stopped pretending to myself that I was anything other than what I was and began to direct all my energy into finishing the only work that mattered to me. Had I really succeeded at anything else, I might never have found

the determination to succeed in the one arena where I believed I truly belonged. I was set free, because my greatest fear had been realized, and I was still alive, and I still had a daughter whom I adored, and I had an old typewriter and a big idea. And so rock bottom became the solid foundation on which I rebuilt my life.

You might never fail on the scale I did, but some failure in life is inevitable. It is impossible to live without failing at something, unless you live so cautiously that you might as well not have lived at all—in which case, you fail by default.

Failure gave me an inner security that I had never attained by passing examinations. Failure taught me things about myself that I could have learned no other way. I discovered that I had a strong will and more discipline than I had suspected; I also found out that I had friends whose value was truly above the price of rubies.

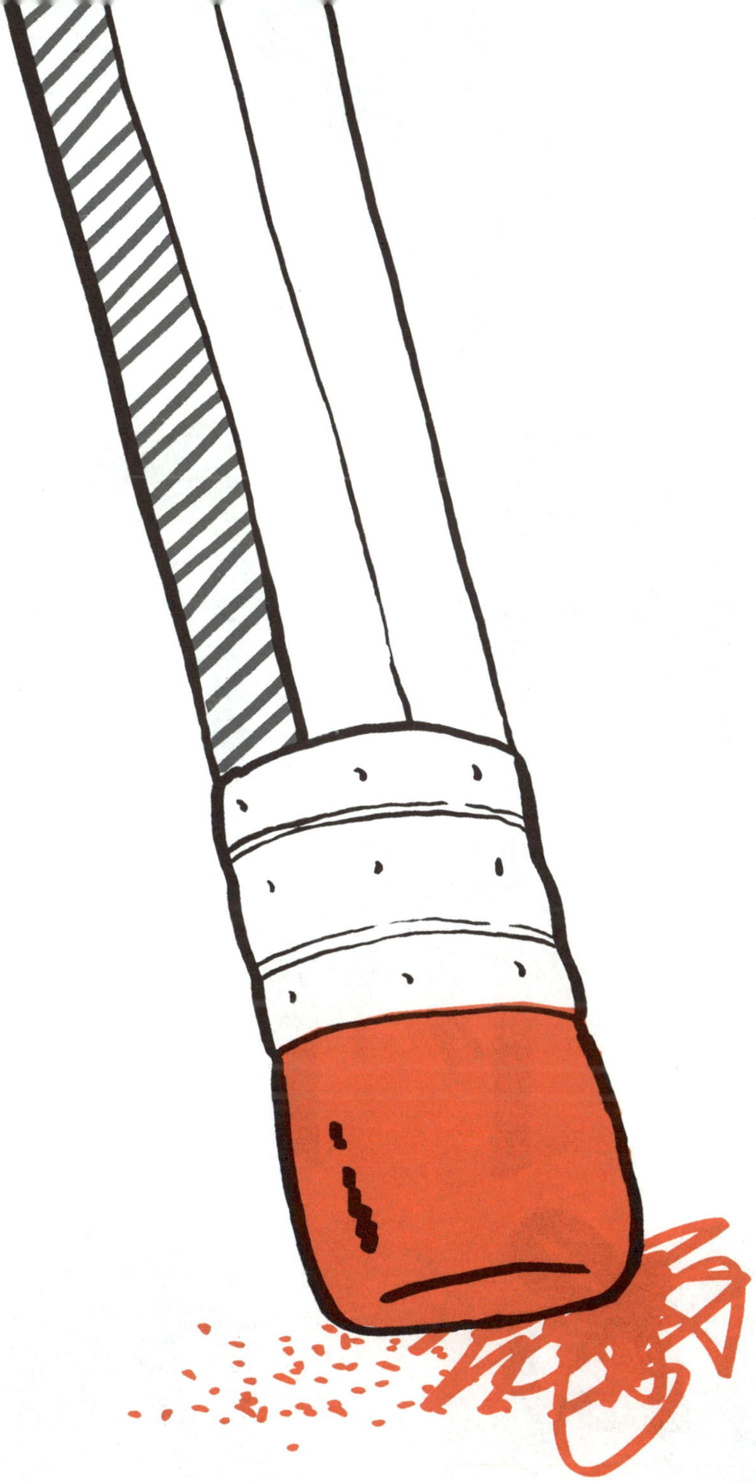

The knowledge that you have emerged wiser and stronger from setbacks means that you are, ever after, secure in your ability to survive. You will never truly know yourself, or the strength of your relationships, until both have been tested by adversity. Such knowledge is a true gift, for all that it is painfully won, and it has been worth more than any qualification I've ever earned.

humility

So given a Time-Turner, I would tell my twenty-one-year-old self that personal happiness lies in knowing that life is not a checklist of acquisition or achievement. Your qualifications, your CV, are not your life, though you will meet many people of my age and older who confuse the two. Life is difficult, and complicated, and beyond anyone's total control, and the humility to know that will enable you to survive its vicissitudes.

Now you might think that I chose my second theme, the importance of imagination, because of the part it played in rebuilding my life, but that is not wholly so. Though I personally will defend the value of bedtime stories to my last gasp, I have learned to value imagination in a much broader sense. Imagination is not only the uniquely human capacity to envision that which is not, and therefore the fount of all invention and innovation; in its arguably most transformative and revelatory capacity, it is the power that enables us to empathize with humans whose experiences we have never shared.

One of the greatest formative experiences of my life preceded Harry Potter, though it informed much of what I subsequently wrote in those books. This revelation came in the form of one of my earliest day jobs. Though I was sloping off to write stories during my lunch hours, I paid the rent in my early twenties by working at the African research department of Amnesty International's headquarters in London.

There in my little office I read hastily scribbled letters smuggled out of totalitarian regimes by men and women who were risking imprisonment to inform the outside world of what was happening to them. I saw photographs of those who had disappeared without a trace, sent to Amnesty by their desperate families and friends. I read the testimony of torture victims and saw pictures of their injuries. I opened handwritten eyewitness accounts of summary trials and executions, of kidnappings and rapes.

Many of my coworkers were ex–political prisoners, people who had been displaced from their homes or fled into exile because they had the temerity to speak against their governments. Visitors to our offices included those who had come to give information, or to try to find out what had happened to those they had left behind.

I shall never forget the African torture victim, a young man no older than I was at the time, who had become mentally ill after all he had endured in his homeland. He trembled uncontrollably as he spoke into a video camera about the brutality inflicted upon him. He was a foot taller

than I was and seemed as fragile as a child. I was given the job of escorting him back to the Underground station afterwards, and this man whose life had been shattered by cruelty took my hand with exquisite courtesy and wished me future happiness.

A
SCREAM
OF
PAIN
AND
HORROR

And as long as I live I shall remember walking along an empty corridor and suddenly hearing, from behind a closed door, a scream of pain and horror such as I have never heard since. The door opened, and the researcher poked out her head and told me to run and make a hot drink for the young man sitting with her. She had just had to give him the news that, in retaliation for his own outspokenness against his country's regime, his mother had been seized and executed.

Every day of my working week in my early twenties, I was reminded how incredibly fortunate I was to live in a country with a democratically elected government, where legal representation and a public trial were the rights of everyone.

Every day, I saw more evidence of the evils humankind will inflict on their fellow humans to gain or maintain power. I began to have nightmares, literal nightmares, about some of the things I saw, heard, and read.

And yet I also learned more about human goodness at Amnesty International than I had ever known before.

Amnesty mobilizes thousands of people who have never been tortured or imprisoned for their beliefs to act on behalf of those who have. The power of human empathy leading to collective action saves lives and frees prisoners. Ordinary people, whose personal well-being and security are assured, join together in huge numbers to save people they do not know and will never meet. My small participation in that process was one of the most humbling and inspiring experiences of my life.

Unlike any other creature on this planet, human beings can learn and understand without having experienced. They can think themselves into other people's places.

Of course, this is a power, like my brand of fictional magic, that is morally neutral. One might use such an ability to manipulate or control just as much as to understand or sympathise.

"They can
think themselves
into other
people's places"

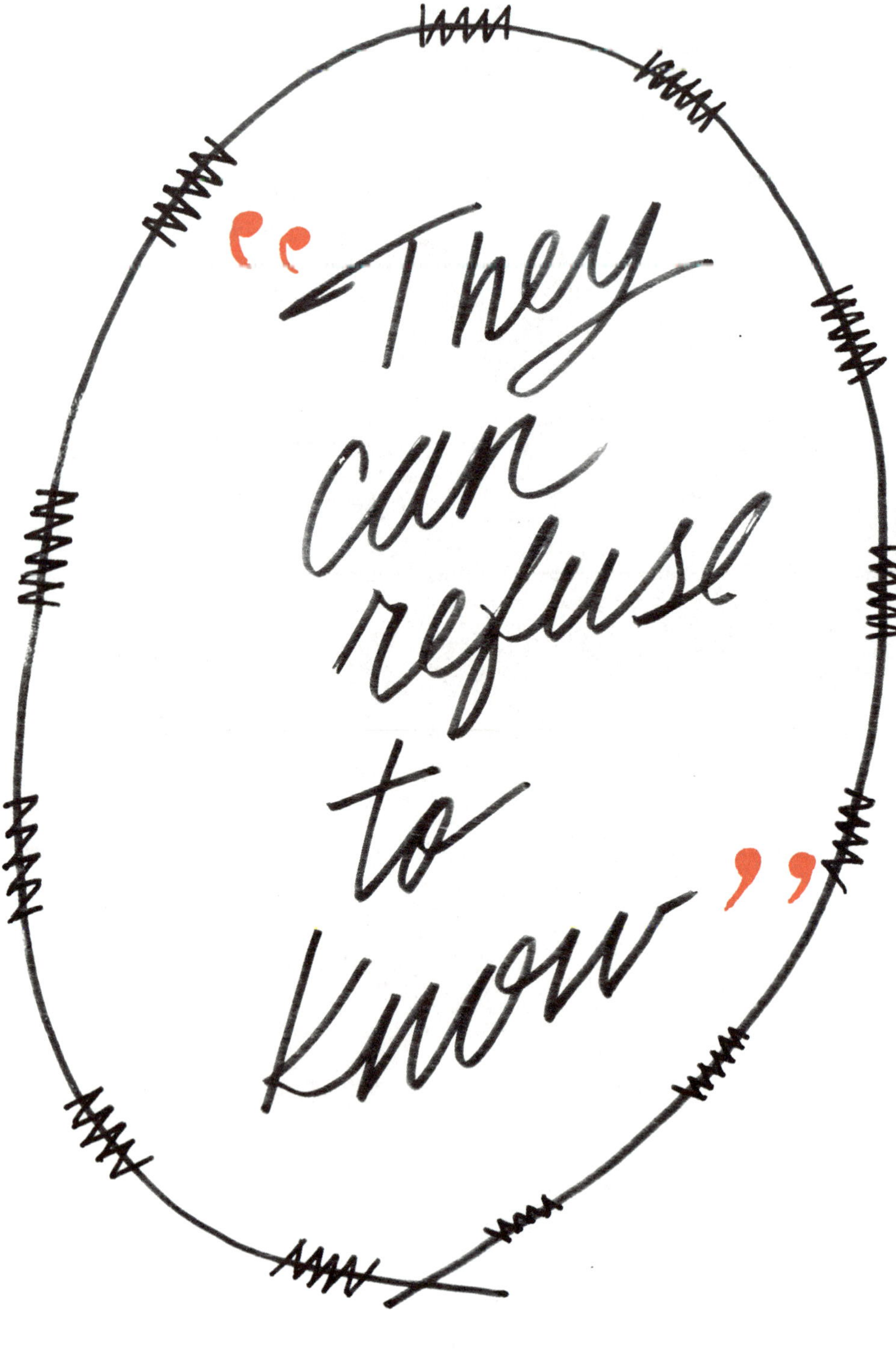
"They can refuse to know"

And many prefer not to exercise their imaginations at all. They choose to remain comfortably within the bounds of their own experience, never troubling to wonder how it would feel to have been born other than they are. They can refuse to hear screams or to peer inside cages; they can close their minds and hearts to any suffering that does not touch them personally; they can refuse to know.

I might be tempted to envy people who can live that way, except that I do not think they have any fewer nightmares than I do. Choosing to live in narrow spaces leads to a form of mental agoraphobia, and that brings its own terrors. I think the willfully

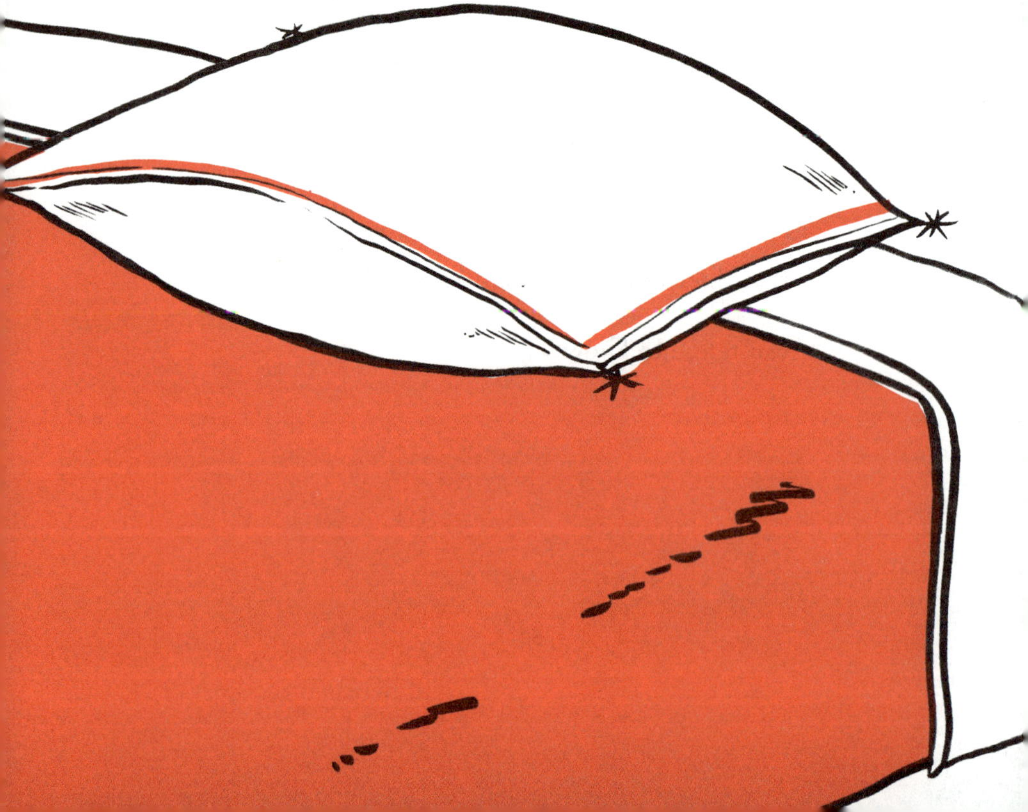

unimaginative see more monsters. They are often more afraid.

What is more, those who choose not to empathize enable real monsters. For without ever committing an act of outright evil ourselves, we collude with it through our own apathy.

One of the many things I learned at the end of that Classics corridor, down which I ventured at the age of eighteen in search of something I could not then define, was this, written by the Greek author Plutarch: 'What we achieve inwardly will change outer reality'.

That is an astonishing statement, and yet proven a thousand times every day of our lives. It expresses, in part, our inescapable connection with the outside world, the fact that we touch other people's lives simply by existing.

N
W
E
S

But how much more are you, Harvard graduates of 2008, likely to touch other people's lives? Your intelligence, your capacity for hard work, the education you have earned and received, give you unique status and unique responsibilities. Even your nationality sets you apart. The great majority of you belong to the world's only remaining superpower. The way you vote, the way you live, the way you protest, the pressure you bring to bear on your government, has an impact way beyond your borders. That is your privilege, and your burden.

If you choose to use your status and influence to raise your voice on behalf of thosc who havc no voice; if you choose to identify not only with the powerful but with the powerless; if you retain the ability to imagine yourself into the lives of those who do not have your advantages, then it will not only be your proud families who celebrate your existence but thousands and millions of people whose reality you have helped change. We do not need magic to transform our world; we carry all the power we need inside ourselves already: we have the power to imagine better.

I am nearly finished. I have one last hope for you, which is something that I already had at twenty-one. The friends with whom I sat on graduation day have been my friends for life. They are my children's godparents, the people to whom I've been able to turn in times of real trouble, people who have been kind enough not to sue me when I took their names for Death Eaters. At our graduation we were bound by enormous affection, by our shared experience of a time that could never come again, and, of course, by the knowledge that we held certain photographic evidence that would be exceptionally valuable if any of us ran for Prime Minister.

So today, I wish you nothing better than similar friendships. And tomorrow, I hope that even if you remember not a single word of mine, you remember those of Seneca, another of those old Romans I met when I fled down the Classics corridor in retreat from career ladders, in search of ancient wisdom:

'As is a tale, so is life: not how long it is, but how good it is, is what matters'.

I wish you all very good lives. Thank you very much.

I wish you all very good lives